BIG CROW PUBLISHING

Published by Big Crow Publishing
Visit phoebefeatherly.com

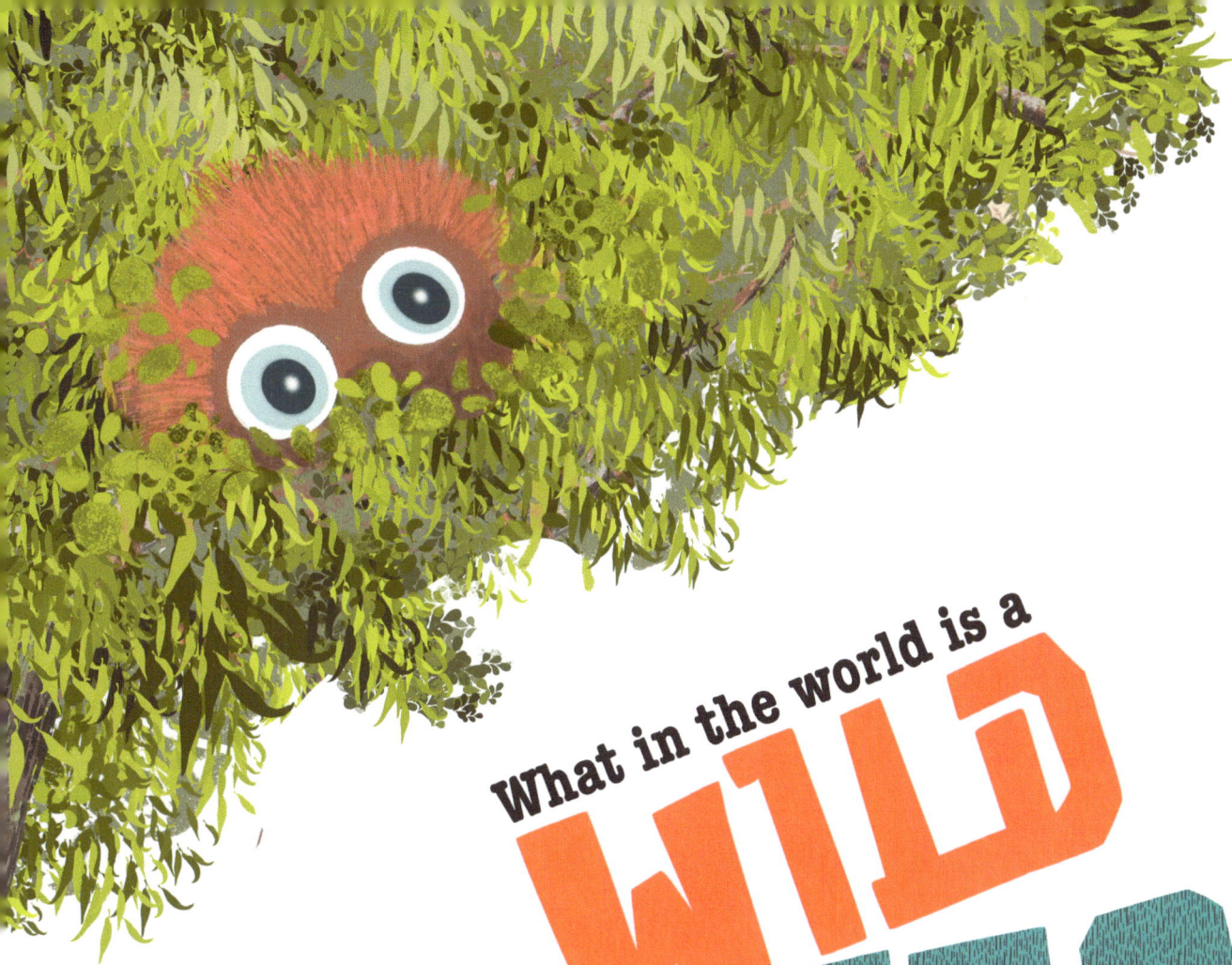

What in the world is a WILD PONGO?

Words and pictures by Phoebe Featherly

To Bernie

(Who actually is a wild pongo)

What in the world is a wild Pongo, anyway? And more importantly, are there any under my bed?

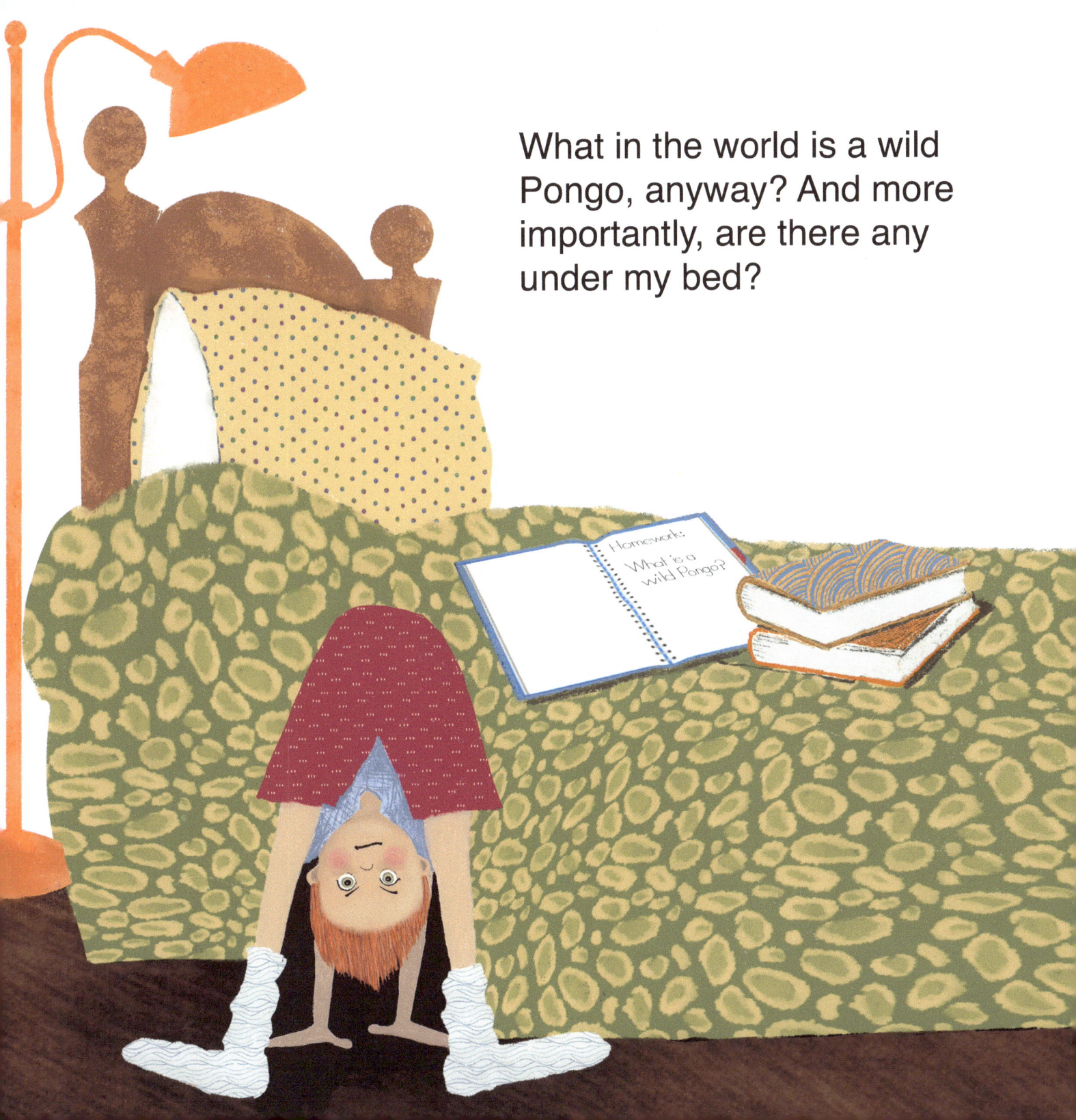

Homework:
What is a
wild Pongo?

My name is Finneas McFarley, and I'm going to the library. I don't know what a Pongo is now, but I can learn a lot more by reading. My teacher said that Pongo is the genus classification of a special animal. She wants me to find their species name!

They only live in the forests of Borneo or Sumatra, two islands in the Pacific Ocean.

Sumatra

Borneo

They share their habitat with monkeys, crocodiles, bearded pigs and sea turtles, just to name a few. They can be hard to see, but there are wild animals everywhere!

This is a bearded pig!

There is a bear, an elephant and a monkey in the jungle. Can you find them?

Adults live alone, but youngsters live with their mom.

Every night, Mom weaves together tree branches to make a nest.

Are they birds?

No, they're not birds. They have long, red hair instead of feathers.

Sometimes, they even make a daytime nest for napping. They love to nap!

They eat fruit and vegetables. Sometimes, as a special treat, they might eat a delicious, juicy bug! Their favorite food is durian. It's prickly on the outside and squishy on the inside. To us humans, it smells like pig poop mixed with gasoline and dirty socks, but to them, it's yummy.

This is a durian!

They swing through the trees instead of walking or running. It's faster and way more fun!

I wish I could swing through the trees!

Can you guess what they look like?

They might weigh as
much as 300 pounds.

That's five of me!

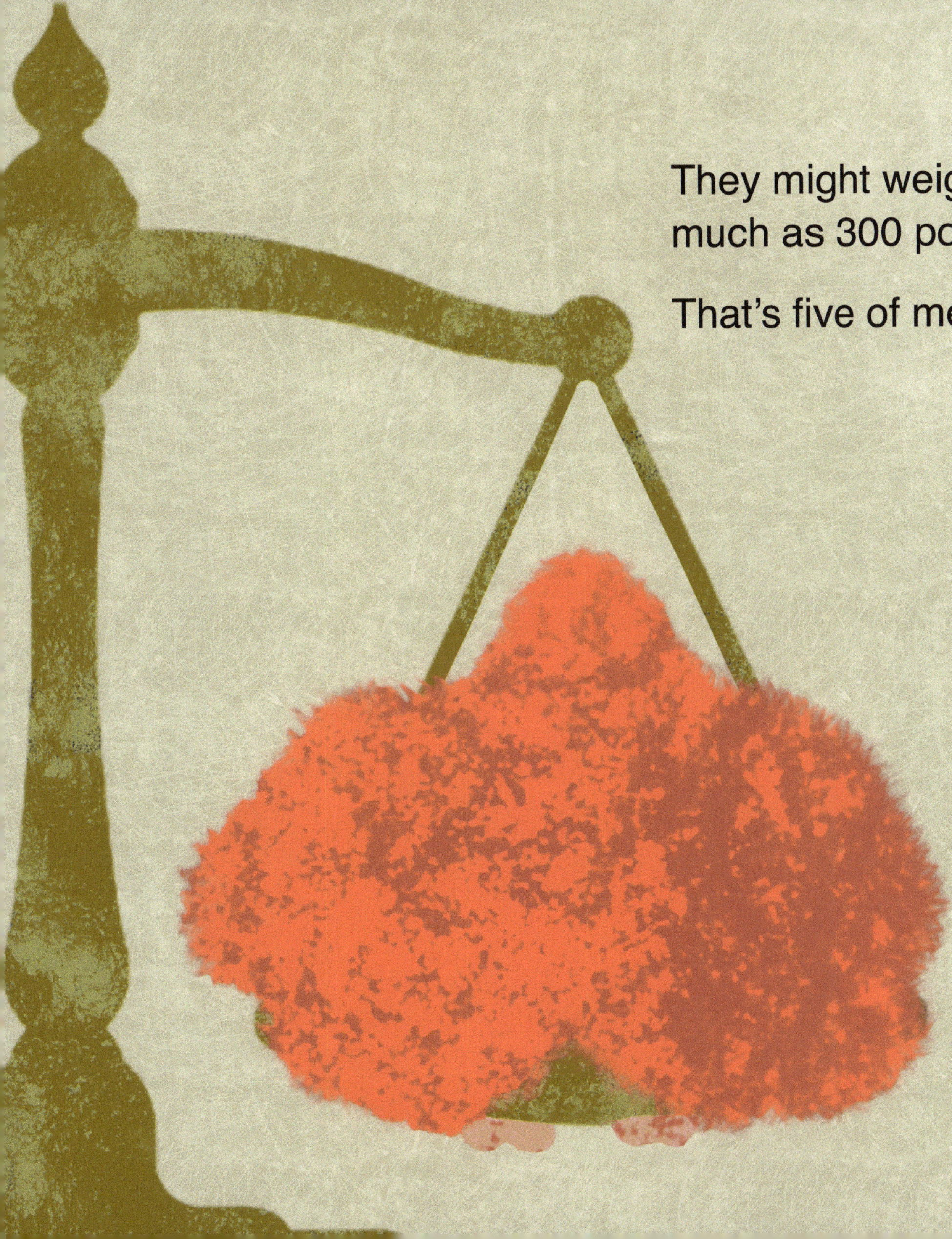

Their legs are short.

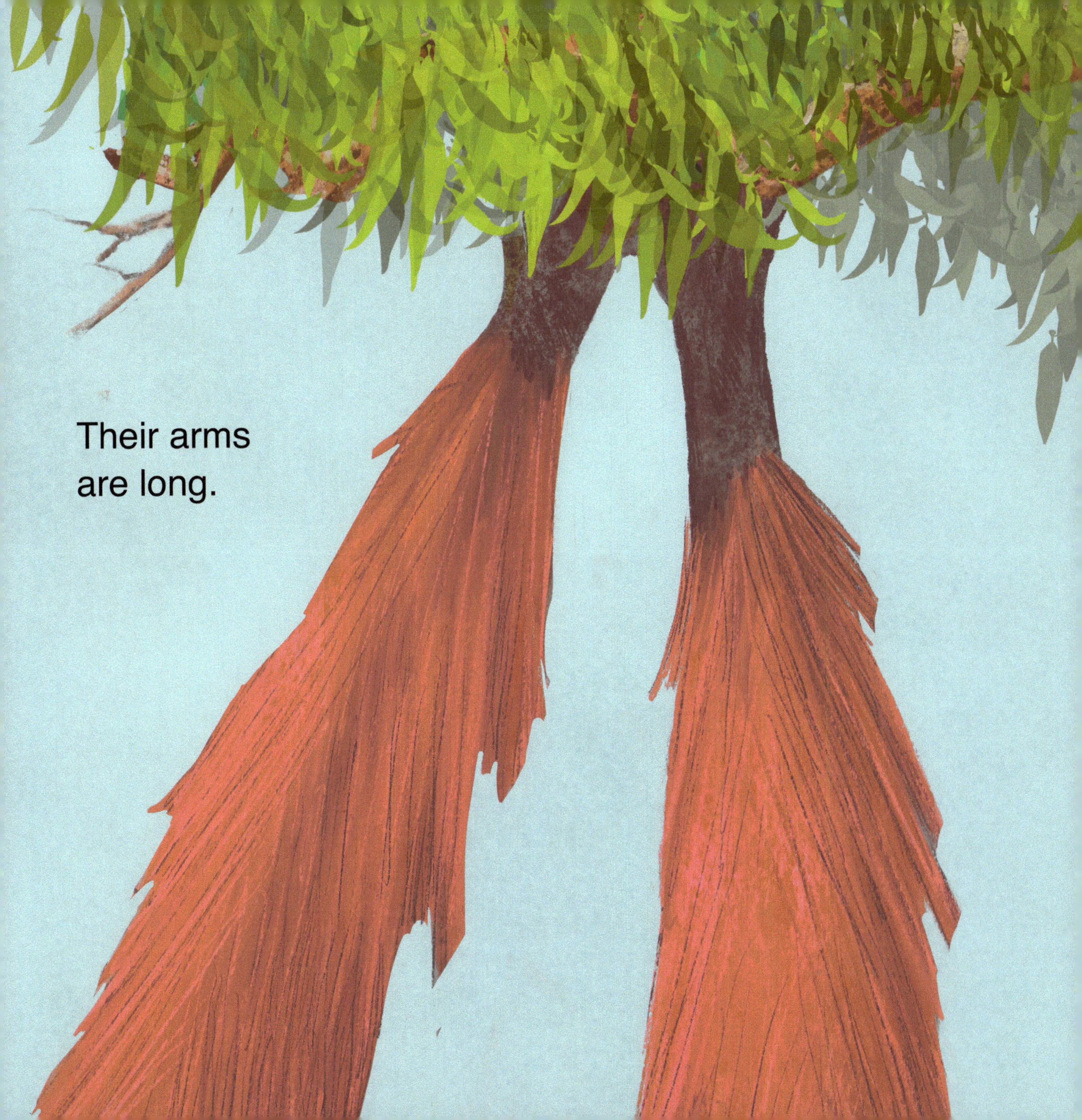

Their arms
are long.

People call them "old men of the forest," but they don't look anything like my grandpa!

Here's another hint . . .

Their common name starts
with the letter O,

but they're neither
oysters nor octopi.
They don't live in the sea;
they live in a tree.

They're not opossums;
remember, they have long
red hair.

They're not ostriches, they have arms, not wings.

They're not otters either; otters don't live in trees.

No, they're not oxen. I don't think you're paying attention. They have red hair, two arms, two legs, two eyes, two ears and no tail.

Can you guess the species name now?

All animals in the genus Pongo are

Orangutans!

Scientists organize all living things into seven parts: kingdom, phylum, class, order, family, genus and species. Here's how orangutans are classified:

Kingdom ~ It's an animal

Phylum ~ With a spinal cord

Class ~ Who is a mammal

Order ~ With hands and feet - eyes face ahead

Family ~ It can do this 👌 with its hands and feet

Genus ~ Pongo

Species ~ Orangutan

Pongo is the genus of three species of orangutans – Pongo bornean, Pongo sumatran and Pongo tapanuli. All orangutans live on the islands of Borneo and Sumatra and look very much alike. Orangutan is a Malay word meaning "person of the forest."

Since orangutans spend their whole lives in trees, rainforests are as necessary for their survival as food or air. The trees need orangutans too. Some tree species can't reproduce until their seeds have gone through the body of an animal. The orangutan eats fruit, swallows the seeds, and poops them out, ready to grow a new tree!

Because some farmers burn the forests to make room for palm oil crops, the rainforests of Borneo and Sumatra are getting smaller. Unfortunately, they are shrinking to the point where there aren't enough trees for the orangutans.

Today, all three species of orangutans are critically endangered and need our help to survive in the wild. You can help by asking your parents to check food labels and try to avoid buying products that use non-sustainable palm oil. If you'd like to help even more, you could ask your parents for permission to start a fundraiser.

There are several organizations with websites where you can learn more about orangutans. Check out the World Wildlife Federation, Orangutan Foundation, and the Sumatran Orangutan Society.

Wild orangutans are in trouble, but if we work quickly, we can stop the destruction of the rainforests they call home. Maybe someday these "people of the forest" will no longer be endangered.

About the Author

Phoebe loves writing and illustrating children's books about how things work here on planet earth. Using fun rhymes and colorful illustrations, she hopes to inspire youngsters to learn more about the many mysteries of science and nature.

Phoebe lives in the forest near the Hood Canal on the Olympic Peninsula in Washington state. She is a retired potter and graphic designer who enjoys kayaking, hiking, working with clay and, of course, reading, writing and drawing!

Visit PhoebeFeatherly.com for printable coloring sheets, games and puzzles.

www.ingramcontent.com/pod-product-compliance
Lightning Source LLC
Chambersburg PA
CBHW060855270326
41934CB00002B/144